"Learning is best done by challenging the old mythologies and this book surely does that."

Prof. Ray Coppinger

TABLE OF CONTENTS

Acknowledgments

My thanks to:

Prof. Ray Coppinger for allowing me to use his photographs, for his invaluable advice and for sharing with me some of his immense knowledge—and for keeping me thinking!

Dr. Ian Dunbar BSc, BVetMed, MRCVS, CPDT for his encouragement, words of wisdom and for his contribution to the book.

Prof. Peter Neville BSc (Hons) and Robert Falconer-Taylor BvetMed, MRCVS, Dip CABT, for their invaluable contributions.

The Centre of Applied Pet Ethology (www.coape.org) which started me questioning the concept of 'dominance.'

Monty Sloan of the Wolf Park, Lafayette, Indiana for the use of some of his wonderful photographs of a wolf (www.wolfphotography. com).

Mrs. Sylvie Derrick and Tarn for raising such a wonderful litter of Border Collies.

My wife, Carol, who had to read, re-read and read again the manuscript and continually corrected my grammar.

And to my old dog Jess who missed her vocation as a canine super-model!

FOREWORD

Dominance in Dogs: Fact or Fiction? is a little book with a big message. Without wasting words, Barry Eaton dispels the dominance myth and its insidious rank-reduction program, which is nothing more than an arduous task for owners to make their poor dogs' lives a misery. "Give them a scalpel and they would dissect a kiss."

The dominance myth 'logic' flow chart is flimsy at the best, but scary at the worst. The notion is that:

1. Wolf social structure is entirely explained by a linear dominance hierarchy in which there is a constant battle to be alpha dog and dominate the rest of the pack.

2. Domestic dogs are descended from wolves and so the same must apply to them.

3. Domestic dogs are trying to dominate us.

4. We should issue a pre-emptive strike and dominate dogs by enforcing strict rules harshly.

In actual fact:

1. Wolf social structure is a wee bit more involved and sophisticated than a single linear hierarchy—this is merely a Mickey Mouse interpretation. Wolves have special friendships and allegiances and by and large, wolves live together harmoniously.

2. Dogs are very (VERY) different from wolves. Domestic dogs were selectively bred for thousands of years to be less fearful and more easily socialized to people. If wolves and dogs were the same, many people would be sharing their homes with wolves.

3. Oh! Get a life!

4. This has to be the flimsiest, most thinly-veiled excuse for little-brained, *schadenfreude* types to label poor dogs as our adversaries in the training arena and in the home.

Why on earth do we treat our best friend like our worse enemy? How on earth can anybody think that a dog is trying to dominate his owners by eating first, going through doorways first, enjoying the comfort of furniture, playing games of tug-of-war, eagerly pulling on leash, or relieving himself in the house? Dogs are not politicians. Dogs are not masters of subtlety or innuendo. Dogs are straightforward and they live in the here and now. If a dog wanted to dominate his owner, he would do just that. End of story. Even so, when dogs bark, growl, snap, lunge, nip or bite, rather than being aggressive or dominant, the dog is usually, understandably, simply fearful of domineering owners.

The 'thinking' behind the dominance myth and the Spartan, boot camp, rank-reduction program is silly to the point of hilarious. Sadly, downright silly thinking becomes extremely serious when dogs are neglected and mistreated as a result. Indeed, many unsuspecting dog owners are bullied by misguided trainers to abuse their dogs under the guise of 'training.'

Certainly, rules are important—any rules—for example sit means sit, and shush means shush. Usually, the owner knows best, especially when the dog's safety is concerned. Also, when dogs and people live together, either we can live with dogs in their doggy dens and adhere to their rules, or dogs can live with us, in our homes and abide by our rules. It is just so much easier for people to teach dogs our household rules and regulations. Moreover, because each dog/human relationship is quite unique, each owner should decide on her household rules for the dog. Each owner should decide where the dog sleeps, for example—on the bed, in the bed, on the bedroom floor, downstairs, on the living room sofa, in a dog bed on the kitchen floor, outside, or in a dog kennel. It is up to each owner to make decisions for her dog. As long as the owner can instruct the dog to lie in his kennel, or to get off the bed, then it's no problem—the dog may sleep wherever the owner wishes.

The greatest joy of living with a dog is being part of creating the most unique inter-specific relationship, wherein "two are halves of one." Enjoy this book. Enjoy your dog.

Dr. Ian Dunbar PhD, Bsc, BVetMed, MRCVS, CPDT

INTRODUCTION

If you only take one fact away after reading this book, I would like it to be the fact that a dog is a dog and *not* a domesticated wolf and, therefore, should *not* be treated as such.

Since the early 1960s, as professional pet dog training developed as an industry, many dog training instructors have based their methods on the theory that an owner must 'dominate' his or her dog, including the use of 'pack rules' to reinforce a dominant position. This was because, back then, many people in the field did indeed view a dog as simply a domesticated wolf. An owner was told he had to train his tame, domesticated dog as though he was a wolf—a theory that was perpetuated by a number of dog training books published during this time. Even today, books, DVDs and television shows based on this theory are still popular and, in my view, cause confusion among pet dog owners as they attempt to ensure their 'dominance' over their pet dogs by being the alpha in their dog/family pack.

The premise that pack rules should apply to modern day pet dogs is incorrect despite the dog being a direct descendent of the wolf. This results in the following *misleading* assumptions:

- A dog's behavior closely mimics that of the wolf.
- Grey wolves form packs based on a structured hierarchy and compete to become alpha by aggressively asserting their dominance over other wolves and, therefore, dogs will do the same.

- Even though the domestic dog lives with humans, he will act as though he is a member of a pack, therefore the owner and family have to be alpha.

Nobody stopped to question this line of thinking until recently. Over the last few years, however, researchers in dog behavior have begun to question the relevance of treating pet dogs as though they are wolves and members of a pack. At the same time, researchers studying wolves in the wild have learned that wolf packs function quite differently than had previously been thought. Many books have now been written (some of which I refer to in the following pages) that dispel the myth that a dog, given the chance, will try to raise his status or become dominant over his human family. Current knowledge and thinking are questioning whether we were ever right to equate a dog's behavior to that of his distant cousin, the wolf.

The purpose of this book is to pull together an overview of the recent arguments that have been put forward by eminent authorities on dogs and wolves as to why the dog is *not* simply a domesticated wolf and, therefore, should *not* be treated as such. It will provide an alternative training and behavioral model to one that insists that the owner must assert his authority over his dog(s) by acting as though he is the dominant pack leader. I believe it's time to open our minds and consider the concept of pack rules as a thing of the past and recognize that dogs are not constantly trying to dominate their owners.

CHAPTER 1

Dominance. What is it?

Mention the word dominance among a group of dog trainers or dog enthusiasts and vigorous debate is likely to break out. At least in part, this is due to both the lack of a commonly accepted definition of the word and to the fact that there are indeed several aspects of the word that we can apply to dogs. So before moving ahead, let's review this hot button of a word as completely as possible.

The human version

According to Dictionary.com, *dominance* as it relates to people is defined as:

1. Rule; control; authority; ascendancy.

In human terms, this can mean anything from being a dictator of a country to rising up the ranks at your place of work. It's human nature to do the best we can in life which may mean gaining a higher status over others. Some people are more ambitious than others so how much a person has control over others is a matter of individual choice and capabilities. We will see later that the alpha wolf is not dictator of a pack, but a benevolent leader, and domestic dogs are not dictatorial and are unlikely to try to raise their status to rule over other dogs in a pack environment.

2. The condition of being dominant.

Your boss at work is more dominant than you because he holds a higher position. A dog trainer standing in the middle of the class is dominant in the sense that she or he has the attention of his/her clients. A boxer winning a fight is more dominant over the guy he's just knocked out. But, all those scenarios can be reversed. One of the dog trainer's clients may be his/her boss at work, so the roles will be reversed when the trainer is at his/her day job. If the two boxers fought again, the one that was previously defeated may win, thus making him more dominant. So, the condition of being dominant may only be temporary. The social structure of a wolf pack will often change. The alpha might stay the same, but others may raise their status, thus others will be lowered. In a multi-dog household, one dog may be dominant over another in a certain situation, but the roles may be reversed in another situation. One dog may be more dominant over food, for example, while the other dog may be more dominant over toys.

3. In psychology: The disposition of an individual to assert control in dealing with others.

This defines dominance in very human terms. We all know people like that, those who always want to be in control or in charge of every situation. People can plan, scheme, volunteer and work to be in control of others. Unfortunately, people are often guilty of applying human values to a dog when analyzing his behavior. Applying such human values raises questions as to a dog's cognitive capabilities. It suggests that a dog has a "theory of mind" or "the ability to adopt the perspective of others" (Udell, Wynne 2008). If a dog can make conscious decisions, he must have some form of language to formulate a plan in his mind. He must be able to plan ahead, not just for the here and now, and be aware of his motives and the consequences of his actions to himself and others. I am not aware of any scientific evidence that a dog *has any of these capabilities*. Such anthropomorphic thinking often leads one to think that a dog has made a *conscious* decision to be conspicuous and influential in the same way certain people are. The flaw in this logic is that people can indeed make such a conscious decision, whereas a dog *cannot* consciously make that choice. Instead, in most cases, dogs develop strategies "based on the greatest chance of reinforcement" (Udell, Wynne 2008). Therefore, this very human definition of 'dominance' would not apply to dogs.

The wolf version

According to Danish ethologist Roger Abrantes (1997), dominance in wolves is defined largely as "a drive directed towards the elimination of competition for a mate." A wolf is driven by a need to breed and pass on his and her genes and will try to assert dominance over certain other wolves of the same gender to enhance its breeding potential. Thus, becoming the alpha wolf increases the odds of producing off-spring.

David Mech (2008) explains that a breeding pair of wolves should be viewed in much the same way as a human family raising their children. The breeding pair's first litter of cubs mature and develop under the guidance of their parents. The following year, when a second litter is produced, the breeding pair are the pack's leaders, and the yearlings are "naturally dominant" over the new arrivals "just as older brothers and sisters in a human family might guide the younger siblings, but still there is no general battle to try to gain pack leadership; that just naturally stays with the original parents." Between the ages of one to three years of age, wolves will leave the pack and try to find a mate to start their own pack. As with the human version above, the applicability of this definition does not readily apply to pet dogs. Given that we humans control (and in many cases, eliminate) when and/or whether our pet dogs can breed, this particular wolf-related definition of dominance also does not apply.

Another issue involving wolf behavior is the common belief that wolves are constantly engaging in fights to maintain or acquire a dominant position within a pack. While fights do occur on occasion, hierarchies are maintained mostly through ritualized behaviors and, therefore, do not involve violence (Abrantes 1997, Mech 1999). Abrantes states, "The hierarchy is defined as a dominance-submissive relationship established and maintained by means of ritualized behavior." These behaviors are communicated largely by a wide assortment of body language signals involving ears, the tail, the eyes and so on. While we humans can learn to recognize many of these signals, we can't raise our hackles and we can't show a similar set of teeth or curl our lips in the same way as a wolf. In short, we are not equipped with the same anatomy as a wolf or a dog to be able to communicate in a way that a dog (or wolf) will fully understand. If we really felt it necessary to be 'alpha' over our dog, how could we establish and maintain by means of *ritualized* behavior in canine language so the dog would understand? The answer is—we can't. We don't have the same ears as a wolf; we don't have a tail. We need to use other tools such as management and training. An example of ritualized behavior can be seen in Figure 1. The wolf on the right is the alpha male; the muzzle at the center is that of the alpha female, now referred to as the breeding pair. The other two wolves are their offspring. This is a 'meet and greet' of pack members using ritualized, genetic canine behavior.

Fig. 1
Courtesy Monty Sloan, Wolf Park, Lafayette, Indiana

Dominance as access to resources

Many forward-thinking authorities on modern pet dogs hypothesize that dominance in our domestic dogs should be more narrowly confined to access to, or control of, *resources*. Resources are anything a dog values including food, water, a warm spot on a rug, toys, even his owner. Dr. Karen Overall (1997) cites Hinde (1967, 1970), Landau (1951), Rowell (1974), and Archer (1988), in defining dominance in this way: "Dominance is a concept found in traditional ethology that pertains to an individual's ability to maintain or regulate access to some resources. It is not to be confused with status." The way a dog can "maintain or regulate" access to the things he values is what is termed as 'resource guarding.' Resource guarding is when a dog acts to maintain or gain control of something he prizes and may show aggressive behavior to keep possession of it. Food bowl guarding is a very common example of this. But, as Overall noted, this activity has nothing to do with status. Once again, 'status' is an example of people wanting to explain a behavioral problem with a concept that applies to humans, not dogs.

It is certainly clear that some dogs are more effective than others at getting access to certain resources. If you have two hungry dogs and put one food bowl down between them, one of the two dogs may consistently be able to gain control over the food and consume it while the other has to wait for leftovers. In that particular situation, one dog may be termed dominant over the other in terms of getting control of the food. However, the results could be reversed if those same two dogs are given one ball to play with. A dog can be dominant in one situation but not necessarily in another. Taking into account research carried out by others, it is my opinion that dominance is not synonymous with a hierarchy or status. Instead, it is the individual's ability to maintain or regulate access to some resource.

So if you want to call a dog 'dominant,' then one should look no further than who controls the resources. This appears to be a more valid concept than 'dominance equals status.' It all boils down to resources and how badly a dog wants to preserve control of, or access to what he values. In the 1970s, biologist Geoff Parker developed a theoretical model called Resource Holding Potential (RHP) that is still used by scientists today. RHP is a mathematical equation that predicts the likelihood of an animal engaging in conflict with another, usually of the same species, in order to gain or retain a resource. The generality of the model is that it applies to all species of animals, including humans. The model assumes that animals do not enter a potentially physically harmful conflict situation before each has assessed (or sized up) the other in terms of ability to win and the value of the desired resource over which the conflict has arisen. It is to the advantage of each individual to have an idea in advance of the likelihood of their winning or losing. The RHP model predicts that a few losses carry more weight in terms of deciding what to do in similar conflicts in the future as does winning.

You can see this happening in dogs that have become trained losers. Even in the most friendly play encounters with other dogs in the park, they roll over on their backs in appeasement possibly resulting in submissive urination. On the other hand, dogs that have learned that using threats of aggression works in getting or keeping what they want may develop an expectation of winning conflicts over re-sources. Clearly, this has absolutely nothing to do with status and everything to do with the resources the dog values most and the strategies he has developed *through learning* to hang on to them.

Dominance aggression

Many people believe that if a dog is showing aggression to his owner he is being dominant and thus, trying to raise his status. Not so. Overall (1997) has a discrete definition of dominance aggression as the "Intensification of any aggressive response from the dog with any passive or active correction or interruption of the dog's behavior or access to the behavior." Simply, this means that if a dog is suffer-ing anxiety due to the behavior that a person or another dog shows toward him, he may become aggressive. Dominance aggression fre-quently results in situations such as when a stranger bends over and tries to pet a dog while he is eating, if a small child corners an anxious dog in a room and then approaches him in an ungainly manner, or in response to an owner who frequently uses punitive training methods or physically punishes the dog for something the owner didn't want him to do.

If a dog feels threatened when facing situations as described above, there are only four coping strategies he can use. These are known as the 4 F's:

1. **Fight:** If the dog chooses an aggressive response and snaps at a person, he or she will most likely quickly

withdraw, if only momentarily. But from the dog's perspective, the snap worked if only for a second, hence reinforcing his behavior. As the dog becomes more successful with his threats, so his confidence grows and he will become more aggressive. The social relationship between the dog and a person he feels is threatening is likely to deteriorate as the dog becomes more aggressive through *associative learning*. (Bradshaw et al. 2009)

2. **Flight:** He can run away, but this will only work until the person or owner catches him. And to make matters worse, he then will probably be punished severely which will intensify his desire to flee.

3. **Freeze:** Dogs (and other animals including people) sometimes employ a 'freeze' strategy, that is, stay completely still in the hope the danger may pass. However, if a person is intent on punishing the dog, this strategy will not work.

4. **Flirt:** This is any type of appeasement behavior a dog might think will defuse or distract from the situation. It might work in some situations, but not against a punishment minded person.

If the signs of appeasement in the flight, freeze or flirt behaviors are ignored and the dog still receives a physical punishment from his owner, his frustration level will rise since such signals are normally recognized and responded to appropriately in the context of dog-to-dog communication. A dog showing submissive behaviors to another dog—who then reacts appropriately—should not have to resort to any aggressive behavior. No wonder the dog at home becomes confused and more fearful when an owner still carries out physical punishment.

Steve Lindsay (2000) states, "Many aggressive displays that are currently diagnosed as dominance aggression are aimed at avoiding some perceived aversive outcome rather than establishing or maintaining the offending dog's social status." And Overall (2003) states that even dogs who are what you might term pushy show "no evidence that (they) are anything other than a variant of a normal dog, and there is no association with any kind of artificial rank hierarchy...and there is no evidence that pushy dogs develop any form of pathological aggression." Rather than being aggressive through some desire to be dominant, Alexandra Semyonova (2006), who has studied a group of dogs through their natural lifetime and in their natural environment, claims that, "In many cases, the inflexibility of human behavior leads to interactions involving aggression. In the first instance, the human is the attacker."

It makes sense, therefore, that a dog may become aggressive when an owner does not understand the concept of resources or can't read the dog's body language, in particular the dog's non-threatening signals. The dog is trying to convey something, but the owner misunderstands and takes inappropriate or no action, or gives mixed signals that only serve to confuse the dog. The dog may also respond aggressively if being mistreated or being harshly corrected. He may be uncertain of the relationship with his owner or with people generally, and possibly react aggressively if something happens that he perceives as a threat. Or, he may be suffering from another form of anxiety that the owner hasn't recognized. This then is a dog reacting, albeit inappropriately from our point of view, to a situation he is not totally happy with. Again, it has *nothing* to do with status.

CHAPTER 2

Wolves and dogs

The idea that one needs to dominate a dog because he descended from the wolf, regardless of how distantly related they now are, has become very ingrained in the world of dog training. Some trainers have lost sight of any other possible way of training a dog. Thankfully that has begun to change. We will examine in this chapter some of the distinctions between wolves and dogs and why a training model or a behavior modification program for pet dogs based on wolf behavior (or what many people think wolf behavior is) is outmoded and ineffective.

When is a wolf not a wolf? When it's a dog

While it's important to understand why a dog is not a wolf, clearly they are related. In fact, most scientists agree that the dog *(Canis familiaris)* is descended from the wolf *(Canis lupus)*. The reasons for this conclusion include:

- The mitochondrial DNA is virtually the same between them.
- They have the same number of chromosomes—78.
- They have the same number of teeth—42.

We also need to acknowledge that there are many behaviors that are shared between the wolf and the dog (digging, circling before laying down, etc.). There is strong evidence therefore that the wolf (Figure 2) is the ancestor of the dog (Figure 3.) However, while the ancestral relationship is clear, it is also evident that a combination of evolutionary processes and selective breeding by man has made the wolf and dog very different animals.

Fig. 2—Canis lupus.
Courtesy Monty Sloan, Wolf Park, Lafayette, Indiana

Fig. 3—Fenn, a domestic dog (Canis familiaris).

It has taken about 14,000 years for dogs of the many breeds and breed types we know and love today to have developed. While the wolf itself has changed very little over this time, we have produced, largely through selective breeding, dogs of all shapes and sizes. The size of a domestic dog can range from a Chihuahua to a St. Bernard. The wolf, by contrast, has stayed within the same size range and has maintained the same coat colors for thousands of years.

There are a number of notable physical changes in the modern dog relative to the wolf. Its teeth have become smaller and more crowded, and the jaws have become weaker. The skull and head shape of dogs are now varied, in sharp contrast to the consistency shown by wolves. Three head shapes of the dog are now recognized (Penman 1994):

- Mesaticephalic: the shape most like that of a wolf, if somewhat smaller, is a medium length muzzle typically seen in German Shepherds, Labradors, and Terriers. About 75% of all dogs have this shape head.

- Brachycephalic: a short, wide muzzle, typically seen in Pug, Pekingese, Shih Tzu, Bulldog and Boxer where the eyes are set toward the front of the skull.

- Dolichocephalic: a long, narrow muzzle typically found in Sighthounds where the eyes are set at the side of the skull.

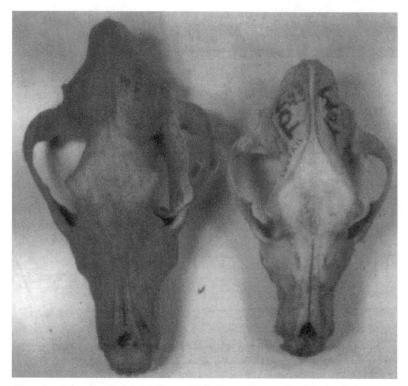

Fig. 4—The skull of a 95 lb. wolf (left) and a 95 lb. dog (right).
Courtesy Ray Coppinger

While the skulls shown in the photo above are from two 95 pound canines, a dog's brain has become about 20% to 25% smaller than that of the wolf, all else being equal. This has lead to the theory that wolves have larger brains and are more intelligent because they have to work to survive, while most dogs do not. The smaller brain in a dog also means that its sense of sight and hearing are less acute than that of a wolf as the need to hunt prey became less critical.

The divergence from wolf to dog began when humans started to build camps and settlements prompting wolves with a short flight distance to remain near people and feed from their village garbage dumps. Food from these dumps provided fewer calories than a wolf's natural prey, but then, dump feeding wolves didn't need to expend energy hunting for food. Instead, they could make better use of the fewer calories in raising their off-spring. Gradually, as the wolf dump feeders adapted to their new niche, they became less nervous, more domestic, naturally tame, and ultimately more trainable (Coppinger and Coppinger 2001).

Other phycsical changes also arose over time. A dog bitch has two oestrous cycles a year and comes into season any time of the year starting between the ages of six and twelve months. Compare that to a wolf bitch that doesn't come into season until she's about two years old and has one oestrous cycle at the same time every year so her pups are born in the spring when it's warm and food is more plentiful (Kreeger 2003) (Packard 2003). A male wolf may not become fertile until twenty-two months of age and then will only be fertile during the annual mating season (Kreeger 2003). A domestic dog, at reaching about six months of age, remains fertile every day of his life.

Domestic dogs go through fewer growth stages than the wolf. They don't fully mature and remain in a developmental stage resembling that of a juvenile wolf throughout their lives (Lindsay 2000). Their ability to communicate effectively through facial and body language and vocalization has changed from the wolf. With the many shapes and sizes of domestic dog we now have, it is not possible for all of them to be able to communicate in the same way. Breeds such as the Siberian Husky and German Shepherd, which most resemble wolf morphology, show more wolf-like behaviors such as dominant/submissive and agonistic behaviors. Breeds such as the Cavalier King Charles Spaniel and French Bulldog show less dominant/submissive and agonistic behaviors, but a high frequency of juvenile play behaviors. The more the domestic dog has diverged from the wolf, the more parts of the body language have been lost (Goodwin et al., 1996). See also Chapter 3.

Impact of breeding

We have bred dogs for hundreds of years for many purposes—for guarding, retrieving, herding, pulling sledges, and hunting. More recently we have bred dogs to provide assistance to people with disabilities and to sniff out drugs, bombs and a plethora of other substances. And, of course, we have bred lapdogs as gentle companions.

A dog's coat can vary in color, length and type, and we even have breeds with no coat at all. For example, the Komondor has a naturally long white coat of matted hair while a Labrador Retriever has short hair that can be sand colored, black or brown. A Chinese Crested has a little patch of coat on his head, tail and paws, but there is no coat on the remainder of his body, so he needs sun block to stop him getting sun burned in summer and man-made coats to keep warm in winter. The Newfoundland has been used for water activities over so many years that they now have webbing between their toes. A few generations ago, Golden Retrievers used to be a Golden brown, but now many are a light, sand color. In some instances, we have bred the same breeds for the show ring and as a working dog. One dog is bred for looks and conformation to meet a breed standard set down by dog fanciers—the other is bred for his working abilities. We, therefore, breed for looks and we breed for behaviors.

We have even changed a dog's gait. At some point during their running action, Sighthounds will have all four paws off the ground, which is in common with the cat family! Wolves always have two paws on the ground when running.

Wolves rarely bark, and when they do it is very subdued. Some dogs bark all the time! In fact, people have intentionally bred breeds of dogs that tend to bark more than other breeds, the Spitz breeds for example. We also have breeds that bark very little, like the Shiba Inu and Basenji.

The dog's brain has changed and not just in size. It no longer thinks like a wolf because it isn't a wolf. A wolf concentrates primarily on three things: (1) hazard avoidance, so it doesn't get killed, or injured and die; (2) hunting, so it can eat; and (3) reproduction, so it can pass on its genes. A dog focuses on what he finds rewarding within his environment. Things like food, toys, walks, companionship, playing with his owner, doing agility, fly-ball, retrieving, herding and all the other things we have introduced into the dog's existence.

The speed at which a wolf's innate behaviors develop also differs from that of a dog. For example, wolves have to learn fear response quicker than a dog—nineteen days for a wolf as opposed to forty nine days for a dog. Wolves have to learn very early on how to hunt and kill in order to survive. Generally, a domestic dog has no need to learn good hunting and killing skills.

A wolf's predatory motor pattern has never changed over many thousands of years. It is still;

Orient > eye-stalk > chase > grab-bite > kill-bite > dissect > eat

During the evolution of domestic dogs, some parts of the sequence of the predatory motor pattern have been enhanced while others have been deleted or are dormant (Coppinger and Coppinger 2001).

A Border Collie's predatory motor pattern for example is:

Orient > **eye-stalk** > **chase** > dissect > consume

Fig. 5—The characteristic 'eye-stalk' part of a Border Collie's motor pattern.
Courtesy Ray Coppinger

Fig. 6—The Border Collie 'chase.'
Courtesy Ray Coppinger

In a Border Collie's motor pattern, the 'eye-stalk' and 'chase' (see Figures 5 and 6) are the most rewarding parts of the sequence. You'll notice 'grab-bite' and 'kill-bite' are missing, but although those motor patterns are still there, for the most part, they are dormant. Unfortunately, when I was participating in sheepdog trials the 'grab-bite' was not altogether dormant in my dog! There were a few trials where I was disqualified when my dog over-enthusiastically grabbed a hind leg of a sheep, but we also won some trials as well!

A Retriever needs to retrieve game with such a 'soft mouth' (see Figure 7) so as not to bruise it, so a Retriever's motor pattern is:

    orient > chase > **grab-bite** > consume

Fig. 7—A Labrador Retriever 'grab-bite.'
Courtesy Mrs CC Guard

The change in predatory motor patterns is further demonstrated by Dr. Erik Zimen (1983). Zimen kept a pack of wolves and a pack of Standard Poodles in adjoining pens and observed their different behaviors. In one experiment, he gave a chicken to one of the wolves and a chicken to one of the Poodles. His observation of the wolf and chicken was that the wolf quickly devoured the chicken, while the Poodle just pulled out feathers and didn't eat the chicken.

In summary

On one hand, we have a wolf whose predatory motor patterns have never changed because they need them intact in order to survive. On the other hand, we have our domestic dogs of different breeds or breed types with dormant or hypertrophied motor patterns that have been influenced by breeders whether of pet/companion dogs or working dogs.

Steven Lindsay (2000) says this about the differences in behavior between the wolf and the domestic dog: "A long history of domestication behaviorally segregates dogs from wolves, and one must take care not to overly generalize between the two canids in terms of their respective motivations and behavior patterns." We still have the wolf *(Canis lupus)* that has remained unchanged for thousands of years, and we have the domestic dog *(Canis familiaris)* that has changed (and continues to change) in looks and behavior and can now do many things that are totally alien to the wolf. While we need to appreciate that although there are many behaviors that are shared between wolf and dog, there are also many behaviors that are totally different. This is why we should not equate a dog's behavior with a wolf's behavior because, quite simply, a dog is not a wolf. As Coppinger and Coppinger (2001) say, "A new species evolves through the gradual shifting, over time, of gene frequency within a population." Dogs are now as far removed from their ancestors as we are from ours.

CHAPTER 3

Packs

Because there is this commonly held notion that dog behavior closely mimics wolf behavior, it is also widely accepted that since wolves live in packs, dogs are also pack animals. So, it is worth taking the time to closely examine how and when wolf packs form as well as how they are structured and function to determine how relevant packs are to our pet dogs.

The wolf pack

Twenty-three wolf experts from North America and Europe have contributed to the book *Wolves: Behavior, Ecology, and Conservation* edited by David L. Mech and Luigi Boitani (2003), both of whom are world authorities on wolf behavior. It is noteworthy that in this entire 448 page tome, the word *'alpha'* is only mentioned six times and then only to explain why the term is outdated. Mech asserts that, "the issue is not merely one of semantics or political correctness. It is one of biological correctness such that the term we use for breeding wolves accurately captures the biological and social role of the animals." (Mech 2008)

Generally, a stable wolf pack consists of a breeding pair and their immediate offspring—essentially a family unit—a fact not generally known until recently. This is known as a 'nuclear family.' However, packs can sometimes consist of an 'extended family,' which include siblings and their offspring; a 'disrupted family,' where one or both of the parents are missing, or a 'step-family,' which has accepted a wolf from another pack (Packard 2003). So pack structure can vary, and social behaviors can be complex, but the point is that these packs are best viewed as families in that they are usually made up largely of related, co-operative animals. In a naturally free pack, the breeding male and female, breed and rear their offspring and usually initiate hunting expeditions. When there are cubs to feed, the breeding female and her pups are dependant on the male and other pack members to provide food. Most wolves at the age of about two or three years will leave the pack and go in search of a mate to start their own pack. These wolves are known as 'dispersers.' Rather than stay in the nuclear family where there is little chance of mating and passing on their genes, these lone wolves will risk the elements of the wild to search for another lone wolf of the opposite sex. If successful, they will then start a new pack, therefore alleviating any social tension in the nuclear pack.

Packs in natural versus captive environments

For a long time, there was a commonly held belief that the wolf pack was held in check by a strict pecking order enforced through violent encounters between the alpha wolves and subordinates. It is now clear that this belief is incorrect, due primarily to the many studies of wolf behavior that were carried out on wolves in captivity, clearly an artificial environment. According to Mech (1999), "in a natural wolf pack, dominance is not manifested as a pecking order and seems to have less significance than the results of studies of captive packs had implied." So, in a free wolf pack, 'alpha' does not necessarily have the same connotation as it does in a captive pack. He goes on to say that "dominance contests between other wolves *are rare* (my emphasis) if they exist at all." Therefore the image of a 'top dog' lording over the rest of the pack is mistaken. As noted above, wolf packs are fundamentally families that function to enhance the survival of the group. Mech says, "The typical wolf pack should be viewed as a family with the adult parents guiding the activities of the group and sharing group leadership in a division of labor....If the kill were small, the breeders would eat first, but if food were scarce, the pups would be fed first. If the kill were big enough, all pack members, regardless of rank, feed together." So, the idea that the alpha wolves always eat first is largely untrue and indeed pups are fed first to ensure their survival.

Much of what we thought we knew about wolf behavior came from the study of captive wolves. Contrary to the family values of a naturally free pack, wolves in a captive pack do make frequent challenges to gain higher status. The higher the position at stake the more vigorously a campaign is conducted. A captive pack will have unacquainted wolves of different ages and gender brought together from different sources. In this situation, a certain amount of social tension is likely to exist, particularly during the mating season. Under these circumstances there is often a dominant male and female, and there are frequent fights among younger wolves for higher status. Within the captive pack, managed and manipulated by man, wolves will be unable to express many of their natural behaviors and will be unable to leave the pack to find a mate as free wolves do.

Unlike the captive pack, a free pack is at risk of injury and death from predators (man, wolves and bears for example). They don't know where their next meal is coming from, so cooperation is paramount. A captive pack lives in an artificial, yet safe environment. They have no threat from predators or injury from large prey and food is plentiful, but there's no opportunity for wolves to leave the pack or to express their natural behaviors. Therefore, the behavior between the two types of packs is quite different.

Packs are formed to enhance chances of survival

As has been shown, wolf packs in the wild are formed in order to enhance the survival of the members and produce offspring. Survival depends on the cooperation of all the wolves in the pack, especially given the need to hunt and kill large prey to supply adequate amounts of food. Even if the pack consists of only two parents and cubs, the cubs need to function cooperatively at a very early age if they are to survive. Can it be then that forming a pack is not entirely genetic but in fact more a strategy for survival? Coppinger and Coppinger (2001), who have studied both dogs and wolves for many years, say this about packing behavior, "Research indicates that packing behavior is a developmental response to a particular habitat. Wolves don't always pack; some populations never pack."

Packing behaviors are emergent. They emerge during the critical socialization period (which is much shorter than a dog's critical socialization period) and depend on social interaction with siblings and older wolves, and the environment in which they are brought up. Different species of wolves around the world will be brought up in different environments and will therefore have different pack behaviors (Coppinger and Coppinger 2007).

Coppinger's hypothesis that packing behavior is not entirely genetic is supported by the work of Schmidt and Mech (2000), "We believe wolves live in packs primarily because adult pairs can then efficiently share with their offspring the surplus food resulting from the pair's predation on large prey." So, if packing behavior in wolves is not entirely genetic, what about pet dogs and even feral dogs?

Feral dogs
Coppinger and Coppinger's research (2001) has shown that even feral dogs do not need to form packs in order to survive. If all the vital elements of survival are available—food, water and shelter—they are happy to live independently or harmoniously in small groups. Figure 8 shows a feral dog in a village in India. The dog doesn't belong to anyone and is not a threat to anyone, but his survival is still dependant on the activities of people. The dog lives in and around the village, so he has the opportunity to scavenge for food discarded by the human occupants, and both water and shelter are available.

Feral dogs live a very different lifestyle to that of wolves. These dogs frequently join and leave the group and there are not the complex rules that wolves live by. The social structure is very loose, whereas a wolf pack is very cohesive. Few, if any, feral dogs are related, unlike the family unit of a wolf pack. When a bitch comes into season, any male can mate with her, and she can mate with more than one male, as there is no breeding pair. The bitch is left to raise her pups entirely on her own, unlike a wolf pack where the entire pack is involved in looking after the cubs. Because of the different social ecology between wolves and feral dogs, Boitani et al. (2008) uses the term 'group' instead of 'pack' to describe feral dogs as they do not fit the true pack model.

Feral dogs have no need to hunt and probably couldn't (apart from small prey such as rodents and rabbits) due to the lack of predatory motor patterns, but all life sustaining resources are available. So, we arrive back at *resources.* Doesn't that ring a bell?

Fig. 8—A lone feral dog wandering around the village.
Courtesy Ray Coppinger

Domestic dogs
We have now established that dogs are not wolves and that the wolf packs formed in the wild are basically cooperative family units with the breeding pair looking after and guiding their young. Yet many people seem to really want to believe that dogs are pack animals and if not kept under strict control will seek dominance over the other members of what we think are a pack. Part of this thinking is probably how we view the world from our own human perspective. Humans live in a culture of hierarchies. From cradle to grave, whatever walk of life, we are almost always answerable to someone. It might seem natural then to pass this hierarchical mindset onto our relationship with our dogs and believe that dogs would perceive themselves as being part of our 'pack.' Therefore, they must have their place within it and to meet with our own perception of the appropriate hierarchical structure, their place should be at the bottom.

While dogs are clearly social animals, do they really engage in pack behaviors? Remember that the foundation of the wolf pack is co-operation to ensure survival. Once wolves lost their fear of humans they started to become domesticated and over time had no need to cooperate for survival. Their behaviors started to change to what we would probably see in feral dogs today. We have seen that feral dogs do not form packs, but instead form groups where dogs join and leave whenever they like. Coppinger and Coppinger (2001) state, "I don't see much in dogs that indicates they have the fundamental behaviors that would allow true wolf-like packing." According to Dr. Ian Dunbar (1979), "In the majority of instances, pack formation in domestic dogs would seem to be an exception rather than the rule." He goes on, "The notion of hierarchies has been much overplayed. For the most part, dogs seem to live in relative harmony with each member of the *group* (my emphasis), each generally going about its business with an apparent disinterest in the affairs of others."

According to the work of Feddersen-Petersen (2008), domestication has changed the dog in so many ways (behaviors and looks) that they may have lost the capabilities of living *exclusively* within a conspecific group (meaning the same species). There is now such a vast range of breeds of different shapes, sizes, coat types, facial and body morphology and breed specific behaviors that the social communication needed to live in groups can be, and often is, confused or misunderstood. Minor disputes can often escalate into major fights as communications will either be missed altogether or just misunderstood causing friction between dogs. Imagine a highly active Border Collie in a play bow, trying to communicate his desire to play with an Old English Sheepdog that has no tail and hair over his eyes. The invitation to play is missed by the OES which could be pretty confusing to the Collie when he doesn't receive a response. This is the antithesis of what goes on in a pack of wild wolves.

Fig 9—A wolf's morphology has not changed for thousands of years.
Courtesy Monty Sloan, Wolf Park, Lafayette, Indiana

Unlike dogs, the wolf has remained the same body shape and size for thousands of years. A wolf can express sixty distinct facial communications, whereas a German Shepherd, morphologically one of the most wolf-like breeds, can only express twelve—and the Pug even less (Feddersen-Petersen 2008). As both body and facial morphology differs greatly between various breeds, it reduces the number of possible expressions and ultimately could result in confusion, miscommunication and an inability to cooperate.

Fig 10—A wolf has sixty distinct facial expressions.
Courtesy Monty Sloan, Wolf Park, Lafayette, Indiana

Fig. 11—A GSD has twelve facial expressions.

Fig 12—A Pug has even fewer than twelve facial expressions.
Courtesy Claire Matthews

So why is it that multiple dog and multiple species households can exist in a peaceful fashion? They can, providing puppies *have been socialized* with other people and dogs at the right age and in the right way (thus the importance of early socialization). If, in the majority of cases, dogs can be socialized to live quite happily together in a social group, why would they feel the need to be part of our 'pack?' In reality, our domestic dog has no reason to form a pack with his human owner as every need for his survival is provided for—by us. And, if the dog has no reason to form a pack with us, there is no reason why we should dominate our dogs or be the alpha in relation to them.

Now, someone is going to say that their domestic dog is in a similar situation to that of a captive wolf. Like the captive wolf, a dog cannot pack his bags and leave. Like a captive wolf, a dog is managed and manipulated by people. However, even though a wolf may be in a captive pack, he is still a wild animal. Release him into the big, wide world and he will still have all his survival instincts intact. He'll still have his predatory motor pattern to be able to hunt, kill, dissect and consume prey; that is he'll still have the *orient > eye-stalk > chase > grab-bite > kill-bite > dissect > eat* motor pattern intact. He'll still have a very strong drive to reproduce. So, what about our domestic dog that has ventured into the big, wide world? Well, how many stray dogs are picked up and eventually find their way into a re-homing center looking more like a skeleton than a dog? They don't have the necessary predatory motor patterns to survive day after day without the help of humans in some way or other.

Domestic dogs, generally, are not equipped to survive in the wild. Just because a dog catches a rabbit one day, it doesn't mean he will be able to survive in the wild. Most dogs will not be able to eat the rabbit because they don't have the 'dissect' part of the predatory motor pattern so they won't be able to get inside the animal where the most nutritious parts are. Even if the dog can dissect and eat a rabbit, he will have to catch rabbits (probably more than one) every day if he is to survive. The energy he expends chasing and catching the rabbit is likely to be more than the energy he receives from eating it—and this is assuming the dog has the speed and dexterity of a rabbit or whatever prey the dog is trying to catch. Wolves are not successful in every hunting trip, far from it, and they are professional hunters working cooperatively as a team. The odds of domesticated dogs, either individually or as a group, hunting for food, surviving without any human intervention, are virtually none.

If domestic dogs have no need or ability to form conspecific packs (and neither do feral dogs), the next question we must ask ourselves is why would a domestic dog perceive itself to be part of a 'pack' with his human family? There's no doubting that dogs are social animals and a dog will certainly form a strong social attachment to his owner and family (and vice-versa). However, during the first few weeks of life (the critical period of social development), the interaction that occurs between a puppy and his siblings and dam results in imprinting of the puppy's brain. In other words, it will know it's a dog, will recognize another dog, and behave accordingly. And, if the puppy is socialized correctly and at the right age, he will be able to live amicably with groups of many different species.

The conclusion seems clear. We accept dogs into our human family, but the dog cannot be part of a pack with humans because of the interaction and imprinting the puppy received in its first few weeks of life. Despite people taking over the rearing of pups from a very early age, the behaviors shown toward people will be dog social behaviors, not human social behaviors. A dog doesn't think like us, or behave like us, or smell like us, or live by the same values as us. John Fisher (1997) wrote, "I really don't believe that dogs look on us as other dogs and, therefore, do not compete with us for status." If we accept the premise that a dog does not perceive himself as part of our human 'pack,' we have to question why he would want to try to raise his status in a hierarchy over humans.

CHAPTER 4

Pack Rules

The widespread misconception that dogs form packs because they are descended from the wolf has led to the notion that dogs form packs, not just with other dogs, but with their human owners, and will try to raise their status over them. Ensuring the dog knows his place in the human pack has led to training systems based on *pack rules*. This concept became fashionable and widespread during the 1980s and 1990s and is still advocated by a number of authors and television personalities.

Depending on which books you've read or what you've been told, pack rules can be numerous and varied. Many dog training classes still give out a list of pack rules which may vary depending on what the instructor has read or learned. An example of the extremes that some trainers will go to invent a pack rule is this one I came across at a training class: "Some dogs get their own way by pretending to be submissive—generally falling on their backs to avoid obeying a command. These are intelligent dominant dogs—don't let them win!" You shouldn't be able to make it up—but somebody has!!

Such pack rule training systems do not take into account either the temperament or character of the dog. We know that some dogs are shy and reserved, yet the harsh regime of living by pack rules is still advocated by some trainers and behaviorists. Just think of the long term psychological effect that this will have on such a dog. Neither does it take into account whether the dog is neutered or not. It's possible that an older, neutered dog that has previously been bred from may still have the instinct to mate and display some behaviors a person might want to extinguish. But surely if someone is going to inflict odious pack rules on a dog, they should at least consider the temperament and physical capabilities/condition of the dog.

Not only are the pack rules used in many a training regime, they are also used for Rank Reduction Programs (RRP) which will be discussed in more detail below. An RRP assumes that any behavior problem is, in most cases, due to the dog attempting to dominate his fellow pack members (other household dogs and humans) by trying to raise his status within the pack. Advocates of pack rule based training claim that almost any problem behavior, whether it is something we find annoying or anything to do with aggression, is a result of an attempt by the dog to gain a higher status. The solution they propose to deal with this is to impose an RRP. Such programs are based to some degree on what people think (largely mistakenly) an alpha wolf would do to keep a lower ranking wolf in his place.

Common pack rules

Let's consider some of the most common pack rules that are proposed as the best way to bring up our dogs, bearing in mind that they are based, supposedly, on how wolves behave and not, in reality, how dogs behave.

Always eat before feeding your dog

This rule is based on the misconception that the alpha wolf eats first. This so-called rule is totally misleading. As mentioned above, alpha wolves *do not* necessarily eat first. Research has found the two breeding wolves eat first *if the prey is small,* but if food is scarce, the pups eat first. If the prey is large enough to feed all the pack, all the wolves eat together at the same time.

A wolf bitch has invested 50% of her genes in her puppies. Her priority is to ensure their survival and she will go without food herself if necessary. Therefore, it's not so much a question of 'dominance' or being 'alpha,' it's more of a question of resources and survival of the young and, therefore, survival of the species. The image of a pack of wolves fighting over a carcass is not found in the wild. Mech (2003) once observed thirteen wolves feeding together, side-by-side, on a freshly killed moose. Only two members of the pack weren't feeding and that was because there was no room around the carcass. Kirsty Peake has been observing wolf behavior at Yellowstone Park, US, since 1999 and lectures on wolf behavior and ecology. In a personal communication (2008), she says she once observed a wolf moving through a resting pack with a piece of meat in its mouth. It squeezed in between two other wolves without any reaction to the presence of food from the other wolves. It's also worth noting that it is not always the alpha that makes the kill, or is even present at the time of the kill, so he won't always be the one to eat first.

Eating something before your dog to show you are supposedly alpha is a rule that does not apply to wolves, so why are we inflicting it on our dogs? And even if wolves did eat first, what would a pet dog actually learn from it? Well, nothing. Imagine it's time to feed the pup. There's dad with twelve-week-old pup's food ready in a bowl placed on top of the kitchen worktop, the pup must wait while dad calls mum in from the garden where she's cutting the grass (I'm all for equality!) then his teenage daughter who is drying her hair while listening to the latest CDs, and lastly his teenage son who is in the garage stripping down his motorbike. They all gather in the kitchen and eat a biscuit. When they've finished the biscuit, they all go back about their business while pup finally gets to eat his meal. Bearing in mind that young pups have four meals a day, this scenario will need to happen four times a day! Also, what is pup going to think when the kids are at school and the bread-winner is at work earning the daily crust? Is this not going to send mixed messages to the pup? Sometimes all the 'pack' is present and sometimes it isn't. What has he really learned from that? Probably that we humans are all mad!

Here is another case. Do all of you trainers who use positive motivating methods, including the use of a clicker and treats, think somehow you are creating a dominant dog? These methods use food as rewards for the dog getting something right. We now have a situation where an instructor and the owner have a bag full of treats for the dog. During the course of the training period, the dog may get to eat all the treats and the owner and instructor eat nothing. Are we making the dog dominant because we are giving him all the food treats and we don't get to eat anything? Of course not, so why should we have to eat something at home before feeding the dog?

Clearly, as Overall (2003) says, "Most training books tell people to feed their dogs after themselves to reinforce the leadership status of the humans. This is wrong."

Fig. 13—A litter of domestic puppies eating first—just as wild wolf pups will do.

Do not allow the dog on the furniture (bed, chair, sofa)

According to pack rules, by letting a dog share your bed, chair or sofa, you are supposedly elevating him to the same or higher status than yourself. If you allow your dog on the furniture at will, you might end up creating a resource guarding problem, but it has nothing to do with status.

The dog may learn that his owner's bed, chair or sofa is a comfortable place to sleep. If he is always allowed access to these resources, he may start to guard them if access is suddenly denied. If you are one of those owners who like their dog to be with them on the sofa or the bed, train the dog to come up by invitation and to get down when told. This will prevent potential resource guarding problems.

Fig. 14—My dog Jess, showing all the signs of not being 'dominant.'

Don't let the dog lay at the top of the stairs

It would appear that this rule has evolved from the idea that the alpha wolf's physical position is always higher than the rest of the pack. Research shows that in a pack with several wolves, the adults tend to lay on higher ground as it's the best position to detect intruders (Mech 2003). As this is defensive in nature and involves more than one wolf, it does not have anything to do with status. Since it appears that the alpha wolf is quite happy to be on the same level as other pack members, that calls into question the validity of the 'do not lay at the top of the stairs' rule.

Dogs have their favorite resting places around the house where they are just content to watch what's going on, chill out or have a nap. A few years ago I had a dog who loved sleeping at the top of the stairs. It was one of her favorite sleeping places because at a certain time of day the sun would shine through a window onto the landing, on the exact spot she was sleeping. Was she being dominant? Nah! Was she being smart? You bet!

Don't let the dog lay in a hallway or in doorways

The idea here is that the alpha wolf positions himself to see the comings and goings of his pack. But since a large pack doesn't always stay together all of the time, it doesn't matter where the alpha positions himself, he won't always be able to see the comings and goings of his pack. However, if the alpha positions himself strategically, this would have more to do with spotting intruders—wolves from another pack, bears and so on—or looking after his mate and cubs.

If a dog positions himself in the hallway, it may just be a resting place. The doorway is a good place for the dog to keep an eye on the owner and be ready to go outside or for a car ride, both things many dogs enjoy. Once again, it shows the dog is smart and knows where the action is. If this were considered a problem, it would probably be one of over-attachment and nothing to do with status.

Never step over the dog

The alpha wolf, supposedly, always makes another wolf move if one is in his way. In fact, research (Abrantes 1997, Mech 2003) suggests that a subordinate will voluntarily move when a higher-ranking wolf enters the 'social space' of the subordinate. The social structure has already been established, so if moving out of the way of a more dominant wolf maintains the social structure, then so be it.

If my dog has found a patch of sunlight to lie down in what happens to be in the middle of the room, I'm quite happy to walk around her. In doing so, I'm not giving off signals of subservience to her and I doubt very much that she perceives that I am. Occasionally, I may need her to move in which case I say "excuse me" because I'm polite and that's what I've trained her to respond to. I'm not exerting 'dominance' over her when I make such a request. Rather when I ask her to do something and she responds, it's because she's been *trained* and for no other reason.

One of the former requirements of Assistance Dogs International member organizations was to teach dogs to lie still while people stepped over them. The standard was changed a few years ago to remove this part of the assessment, not because it was a dominance or status issue, but because they feared the dog might sustain injury in public if it stayed lying down when someone stepped over it. Canine Partners continue to train their assistance dogs to lie still, but also train the dog to move when asked to do so (Bondarenko 2007, personal communication).

Making a dog move won't make it more subservient or enforce the owner's dominance. There could be good reasons for wanting the dog to move or to lie still, but this is achieved by training, resulting in a well-trained dog, not a subservient one.

Never let your dogs through a doorway first

This we are told is because it's the alpha's privilege to go first. This rule has clearly come from observations of a captive wolf pack—there are no doorways in the wild—where the alpha may well go through small openings first, for example when transferring from one pen to another. Clearly this is not some genetically based behavior that a pet dog might have inherited.

Even *if* owners follow this flawed rule, subservient wolves will show signs of deference as alpha goes first. People cannot mimic the posturing of an alpha (breeding) wolf, and a dog won't show a submissive posture as the owner goes though a door first, so the entire exercise is pointless. It means nothing to the dog and the owner achieves nothing.

When it's pouring rain and the dog has to go outside to eliminate, would you go out the back door first and get soaking wet just to enforce your alpha position? I certainly wouldn't. My dogs will have to go out on their own and get wet. But by doing that, I won't be telling them that they are of a higher status than me because I can't give the right canine signals. And what would happen if you were sitting watching television with your dog lying by your side, and then he gets up and starts wandering towards an open doorway? Would you have to get up and dash to the door to make sure you go through first? And what if you have more than one dog and they're in different parts of the house? There are so many permutations of different scenarios, that to enforce this 'rule' becomes impossible—and unnecessary.

The one occasion I do agree with an owner going through the doorway first is when taking the dog (or dogs) for a walk. The last thing you would want is your dog pulling you through the front doorway in his excitement to get to the park, but this is just good manners and safety based on training, not showing who is being dominant.

Never let your dog pull on a leash

The reasoning behind this rule is based on the misconception that the alpha wolf in the wild leads the way and dictates where the pack goes. The alpha may decide on the route to take, but does not always lead from the front. According to Mech (2000), leading a pack can be influenced by "youthful exuberance and oestrus." He continues on by saying that "Wolves often follow river beds, game trails and old roads. When doing so, it is obvious where the pack is headed for certain stretches, so any wolf may forge ahead temporarily."

Mech also describes how he observed a pack of wolves crossing a frozen river led by the alpha. Most of the pack started to cross while some wolves were hesitant and stayed on the river bank. Part way across other wolves felt uncertain and turned back. Eventually the rest of the pack turned back. The pack did not follow the alpha, but instead made a cooperative decision as to which way to go.

Contrary to the reasoning behind this rule, the alpha wolf does not always lead first, so to say a dog is pulling on the lead because he's trying to exert dominance is totally misleading. Dogs pull on the lead because they haven't been trained not to, or are excited to get to the park, not because they are seeking a higher status. If we did follow this misguided rule, surely we should be teaching our dog to walk behind us, not next to us! In my opinion, this rule, like all the others, is a convenient way to explain to owners why they are having problems with their dog. Let's blame the wolf! A realistic scenario is a that a dog loves going to the park, and in his excitement he pulls on the lead in the hope of getting there quicker. On the way home, when the dog is tired, he walks nicely on a loose lead by the owner's side. In the meantime, the dog has been running free, often ahead of the owner. But that does not mean the dog is being dominant on the way to the park and while running in the park, but is being subservient on the way home. I don't think that's very likely.

Never let your dog initiate play or seek attention

Another interesting, but flawed, pack rule is that since the alpha wolf initiates (supposedly) the start and end of play and demands attention, we should not let our dogs do the same. What we know about a free roaming pack of wolves is that adult wolves of both sexes care for and show tolerance within the family. A wolf pack needs strong social bonds. According to Mech (2003), "The psychological tendency to form (strong) bonds results from a mere desire for physical contact. As pups grow older, physical contact continues during play and eventually occurs daily among all members of the pack." Also, "any highly motivated wolf can affect the activity of its pack mates, such as play." Zimen (1981) says, "No member decides alone when an activity is to begin or end." So, it doesn't appear that a pack of wolves sit around all day waiting for the alpha to initiate a play or seek the attention of others.

The bond between domestic dog and owner must also be strong if they are to co-habit harmoniously, so why can't a dog initiate play or seek attention? Dogs are social animals and need social contact. Some dogs, however, will take advantage of being given too much attention and will start to *demand* it. If owners are inconsistent in whether they give attention or not, it may result in a confused dog developing unwanted behaviors such as barking or jumping up in order to get the attention it desires. To avoid confusing the dog, owners must be consistent in their actions and train the dog. Dogs need to learn good manners and like everything else, this comes through training. It has nothing to do with status or pack rules.

Do not play games of tug and never let your dog win

This rule refers to wolves tugging on a piece of meat where, in theory, the higher status wolf would always win. However, in the wild, wolves tug on meat for a very practical reason. Wolves will open and dissect the tougher parts of a carcass by grabbing an end of some part of the prey's anatomy and tugging against each other to make it easier to tear the skin apart or to pull muscle meat from bones. Each wolf eats whatever he ends up with. As in so many facets of wolf behavior, tugging on meat is not dominance related, it is a cooperative effort for the benefit of the pack. This erroneous link to wolf behavior has led some people to claim that the dog, if allowed to win a game of tug, would perceive itself as stronger than its owner and could lead to dominance problems.

Research into the differences between dog-dog play and dog-human play at Southampton University (Rooney, Bradshaw and Robinson

2000) found that dog-dog play is more of a contest. Dogs behave differently when playing with other dogs than they do when playing with humans as dog-dog play is more competitive than dog-human play. No evidence could be found that dog-human play was anything more than a game, not a contest. The research concludes that, "Decreased competitiveness may mean that the outcome of dog–human games is less likely to affect the players' relationship than has been suggested by some authors."

To avoid any possible problem of resource guarding of a favorite toy, train the dog to 'leave it' or 'drop it.' Play is important for learning, influencing behavior and forming a bond between dog and owner. Tugging is a natural behavior and we have enhanced that behavior to serve man. For example, an assistance dog will tug the washing from the washing machine, but it doesn't make him dominant. There is nothing wrong with playing tug with a dog and even letting him win sometimes, providing you have taught the dog some etiquette like responding to a 'drop it' command.

Stand in your dog's bed to show him you are alpha

The alpha wolf is supposed to be able to sleep where he likes and will make another wolf move if he fancies his sleeping place, so we are supposed to stand in the dog's bed to show that we are alpha.

In reality, for the first few weeks, wolf cubs cuddle up together, but from about four weeks of age they develop social distance and from then on sleep apart, including the alpha. "Contact between sleeping animals is rare and occurs mostly by chance." (Zimen 1981). Also, Mech (2003) says about this wolf sleeping arrangements "If allowed to choose their own resting sites, they usually select separate ones." In a personal communication, Peake (2008) says that over the many years she has been observing wolves, she has never seen an alpha move another wolf just to gain its sleeping place. Her colleague at Yellowstone Park says, "All the times I've watched them there doesn't appear to be any indecision or bickering over bedding spots."

Looking at the scenario realistically, we stand in the dog's bed, we then get out, and the dog gets in. What is the point? What is the dog going to learn from that? To exert our authority over our dog, the pack rule says if the dog tries to get in the bed while we are standing in it, we are told to take the bed away and make him sleep on the floor. Not all homes have fitted carpets. Some have a polished wooden floor and others may have a stone floor in the kitchen where the dog would possibly sleep. While these features may be aesthetically and architecturally appealing and enhances the character of the house, they don't make very comfortable sleeping surfaces. To deny a dog comfort and warmth is tantamount to cruelty. And just as a thought, what if the dog's bed is a crate? Or, maybe you have a small dog that likes to make his bed in unusual places (see Figure 15)!

Fig. 15—What purpose would standing in this dog's bed serve?

Put your dog in a "down" position to enforce your status as alpha

I recently watched a commercial video of a dog trainer instructing a class over a six week period. All the pack rules were mentioned including the importance of putting your dog in a down/stay position. The owners were told to teach their dog a thirty minute down/stay just to show their dog that he is subordinate to his owner. What a miserable life those dogs must have. These are pet dogs with novice owners, yet they had to teach their dogs a thirty minute down/stay. In competitive obedience at championship level, dogs are only expected to do a ten minute down/stay, but the poor dogs in the training class were expected to go for thirty minutes, just so the owners can show their dominance over the dog.

Figure 16 shows my dog, Jess. I have just asked her to lay down, and she has. So, I say "good girl" because I've asked her to do something and she obeyed. Then I give myself a pat on the back for being such a good dog trainer! But does she look submissive or subordinate? On the contrary, she looks perky and ready to go!

Fig. 16—A well trained dog!

The alpha roll

This last rule results in using force to physically put the dog on his back and into a submissive position. This is called an *alpha roll,* an action that is thought to mimic wolf behavior. When a dog is subjected to an alpha roll by a person, it usually involves grabbing the dog and rolling him on to his back, then pressing him down while shouting at him until he submits. What actually happens with wolves is that a subservient wolf *voluntarily* adopts a submissive position on his back in reaction to signals displayed by a higher ranking wolf.

Harrington and Asa (2003) assert that, "Passive submission is often a reaction to approach and investigation by a dominant animal. The submissive animal lies partly on its back, with its tail curved between its legs and its ears flat and directed backwards." At no point is a wolf or dog *forced* into an alpha roll in order to submit.

In terms of dog behavior, Patricia McConnell (2002) says that, "Well-socialized, healthy dogs don't pin other dogs to the ground. Submissive individuals initiate that posture themselves. The posture is a display signal from one animal to another, a signal of appeasement, not a wrestling maneuver. Forcing dogs into submission and screaming in their face is a great way to elicit defensive aggression. Within their social framework, you're acting like a lunatic."

In the real world of dog–to–dog communication, the Spaniel in Fig. 17 is showing submissive behaviors toward the more assertive Collie. The Spaniel then turns on his side exposing genitalia. The Collie then investigates the Spaniel (Figs 18 and 19), sniffing head and tail, the prime areas for hormones, to find out as much information as he can: sex, age, neutered, etc.

When the Collie has completed checking out the Spaniel, the Collie backs off and the Spaniel moves away (Fig. 20). There has been no aggressive behavior by either dog. The Collie does not force the Spaniel to the ground and bark in his face. Yet that is what is advised by people who believe alpha rolls will make a dog submissive. When an owner performs an alpha roll on their dog, he is being confrontational. The dog can either resign himself to the situation, or he may not like it and nip the owner and that may escalate into 'dominance aggression' which was mentioned in Chapter 1.

Making a dog submissive is not something you can train a dog to do. "Be submissive!" No, it doesn't work like that. A submissive behavior is innate. It's a natural behavior, it's part of dominance/submissive ritualized behavior that is hard wired in a dog.

Fig. 17

Fig. 18

Fig. 19

Fig. 20

Pack rules—the end

If we are going to transpose wolf behavior to our dogs, which I believe is totally wrong bearing in mind how the domestic dog has diversified so much from the wolf, then let's at least get the wolf behavior right. The alpha doesn't always eat first, lead the pack, occupy the highest ground, and initiate the start of games or attention. In fact, all the rules mentioned above are flawed to the point of being ridiculous. They don't apply to wolves and they don't apply to our dogs.

Comparing how wolves behave and then transposing that behavior into a set of rules as to how you should be treating your dog just doesn't work. The dog won't understand what you are trying to do or what message you are trying to convey, so you'll end up with one totally confused dog because you don't have the anatomy or the innate dominant/submissive behaviors to communicate in canine language that the dog will understand.

In its heyday, pack rules were seen as the answer to all behavioral problems, even to some that didn't exist! Some people followed the rules too rigidly and ended up with a fairly miserable dog. Many behavior counsellors would, and some still do, recommend pack rules to solve a behavior problems. John Fisher was a canine behavior counsellor, renowned lecturer and excellent trainer. He wrote about owners who wanted to be alpha (1997), and said of them: "If it's how you want to live with your dog, I have news that is going to disappoint a lot of people who have striven to reach this Alpha status—it all means diddly squat to your dog!"

There is further testament to this fact from McConnell (2002) when she says, "Wolves do a lot of things that we have no reason to emulate, from eating the placenta of their newborn to killing visitors from other packs, so recommending that we humans should do something simply because wolves do is not a compelling argument. Dogs are not behavioral replicates of wolves."

If you can accept that:

- Domestic dogs are not wolves, and neither do they act as if they are.

- Pack rules are flawed to the point that it's questionable whether they apply to free roaming wolves, let alone dogs.

- The human family is not a surrogate pack for the dog, but the dog is part of our social unit.

Then maybe it's time to put an end to applying the outdated, hand-me-down 'pack rules' to our dogs.

Rank reduction programs

When a dog develops a behavioral problem, it's just that—a behavior problem. Too often this simple fact is ignored in the belief that it is a 'dominance' problem. A Rank Reduction Program (RRP) is a so-called behavioral cure-all for supposedly dominant dogs which focus on teaching the dog that he belongs at the bottom of the family hierarchy—but it does not take into account dealing with *specific* behavioral problems. It is a harsh program that enforces pack rules and denies certain types of rewards and pleasurable activities, the combination of which is likely to have a devastating effect on a dog's daily life—particularly if he has not been subjected to these conditions before.

In a typical RRP, a misbehaving dog who has been used to getting strokes or cuddles from his owner will now be ignored. A dog that previously asked to be played with by dropping a ball in his owner's lap will also be ignored. A dog that is happy to play by himself with a toy, or chew on a chew toy, will no longer be able to do so. A dog will no longer be allowed to join his owner on the couch, even if he has previously been invited to do so. A favorite sleeping spot may also be denied to him if it's on top of the stairs or near a door. The dog receives no rewards from his owner for good behavior and will also be denied *life rewards,* those things that have become important parts of the dog's daily routine that he enjoys and finds rewarding

Rank Reduction Programs are ineffective and psychologically cruel. Neither wolves nor dogs use an RRP and *neither should we.* Effectively, it means the dog's life is going to be turned upside down because his normal expected daily life rewards will be denied him as the owner attempts to assert his dominance over the dog. According to Fisher and Whitehead (2001), "If you remove an expected reward, you are in all aspects other than physical, punishing the dog." The result of being denied expected life rewards and therefore being randomly punished, "could cause conflict, depression, response suppression and even helplessness." The strokes and cuddles, the games, food treats and all the attention he receives on a daily basis will no longer be part of his daily life as all the pack rules are imposed which deny the dog such daily rewards. This regime of mental cruelty could suppress the unwanted behavior, but what happens when the owner believes the problem has been cured and life returns to normal? The unwanted behavior is likely to return because the *specific behavior* has not been addressed.

Given the wide variety of breed-specific behaviors, dogs will respond to the harsh regime of an RRP in different ways. Imposing an RRP is likely to cause the dog psychological harm and he will attempt to find emotional relief to the frustration caused by the denial of resources that are most important to him in one of two ways. A dog might try harder to gain access to the resource he is being denied which may well involve the dog using aggression to get the resource. This, obviously, is going to make the situation worse. Alternatively, the dog may resign himself to the situation, resulting in depression. In this second case, it can appear to the owner that the dog has improved and the problem has been cured, but in reality it hasn't.

If we look at these two possibilities of response to an RRP, think about how a Labrador is likely to react if he is made to sit and wait for a meal or denied some other significant reward that he may have enjoyed on a daily basis, such as retrieving a ball during a game with his owner. How would he feel? How quickly might his restrained frustration evolve into anger, or how long before he might give up waiting and find something else to do? Perhaps quite a while because a Labrador is likely to endure the frustration better than many breeds. If the RRP carries on for too long, the Labrador may give up trying to gain one of his prized resources which will result in *learned helplessness* due to the RRP. Symptoms of learned helplessness are the inability to experience pleasurable emotions from normally pleasurable life events such as eating, exercise or social interaction; and is charactarized by depression and the lack of motivation, emotions and cognition (McGreevy, McLean 2009).

Now think about how a Jack Russell or German Shepherd might react if they are made to sit and wait for their meal or denied the chance to play games with their owner. Their endurance of frustration will probably be a lot shorter period of time than the Labrador. Their frustration may result in aggression in order to gain the resource that's being denied, hence making the behavioral situation worse. So, the dog's behavior has worsened resulting in *learned aggression* because of the RRP. Learned aggression is when a dog has used aggressive behavior to avoid unpleasant stimuli and found that it had the desired effect, thus reinforcing the behavior (Bradshaw et al. 2009).

Some people use what is called a Nothing in Life is Free (NILF) program to train behaviors. A very typical one might involve getting a dog to 'sit' before he is let out the door. Being let out the door is the reward for the sit behavior. However if a NILF program is taken to extremes, it means a dog cannot do anything he finds rewarding without having to 'work' for it. He can't even take a drink of water without having to do something to earn the drink. While used in moderation, it may have its place in training and teaching good manners, but it will not resolve more serious problems such as a couch-guarding dog.

An RRP (or an extreme NILF) cannot possibly affect all dogs of so many breeds with differing breed-specific behaviors, characters and temperaments in the same way—not that it would have any good effect in the first place. Clearly in each case the individual dog's needs, personality and emotions must be assessed. *A specific unwanted behavior must be tackled directly through a specific individual approach to have the best chance of success.* This comes through teaching and rewarding the dog to behave differently in certain problematic circumstances, not through the blanket cure-all of a Rank Reduction Program.

Do rank reduction programs work?

There will always be somebody who has used a Rank Reduction Program and said it has worked. According to John Fisher, random punishment *suppresses* behavior and by denying the dog his daily rewards, the dog is being randomly punished. Therefore, certain behaviors are suppressed and it may indeed be the behavior the owners are trying to eradicate is among those that are suppressed. But at what cost to the dog psychologically? And what happens when the owners think they've overcome the problem and life returns to normal? Will the unwanted behavior re-appear? Quite possibly it will, because the specific problem has not been addressed.

As an example, if we take a couch-loving dog who is showing aggressive behavior when the owner tries to move him, is he displaying dominance in that he is trying to raise his status? No. Is he guarding a resource? Yes. Therefore there is a singular behavioral problem the owner needs to modify, and it is this *singular problem* that must be addressed directly. An RRP does *not* treat a singular problem.

I cannot stress enough that if a dog has a behavioral problem, whether it be aggression or just something annoying, the *specific* problem must be addressed. There is no cure-all, so forget pack rules and concentrate on the problem. Indeed a RRP is a pointless exercise which can be tantamount to physical and psychological cruelty if taken too far. As Jean Donaldson (1996), says, "The whole dominance idea is so out of proportion that entire schools of training are based on the premise that if you can just exert adequate dominance over the dog, everything else falls into place. This is dangerous." And she goes on to say it means "that incredible amounts of abuse are going to be perpetrated against any given dog." This is echoed by Karen Pryor (2002) who says, "...perhaps only we humans use punishment to gain for ourselves the reward of being dominant." She goes on to say that if you want a dog to change his behavior, "it's a training problem and you need to be aware of the weaknesses of punishment as a training device."

CHAPTER 5

A definition of dominance that makes sense

Having accepted a dog for what he is—a domesticated, tame animal with different behaviors and motivations than a wolf—we may be in a better position to consider objectively a different, more up-to-date hypothesis of dominance based on science and what we now know about wolf and dog behavior. The use of the term or concept of dominance in a dog-dog or dog-human relationship should be limited to: (1) the ability of a dog to regulate access to and retain resources; or (2) the degree to which he engages in dominance aggression or a combination of the two. Refer back to Chapter 1 for introductions to these concepts.

Resource guarding

Earlier, I quoted Overall who says, "Dominance is a concept found in traditional ethology that pertains to an individual's ability to maintain or regulate access to some resources. It is not to do with status." If we limit the definition of dominance in dogs to *the ability to maintain or regulate access to resources*—and exclude concepts such as packs and status—we now have a concept that actually describes behaviors we see in our pet dogs and, through proper training and management techniques, can usually be controlled successfully by owners. Dominance is more a question of winning or losing access to resources, not about gaining higher status. Dogs who have *learned* that using threats of aggression works in getting or keeping the resources they want may develop an expectation of winning conflicts over resources. This would explain why some dogs may become aggressive when an owner tries to remove them from his couch or favorite chair. It may explain why some dogs guard their food or toys. If this is the case, then the relationship between dog and owner should be reviewed and changes made.

Food guarding is a common problem in which the dog becomes labeled as dominant. There are two main reasons why this occurs and both are totally preventable. The first is the puppy who is left on his own to eat his meals in peace and continues to be left undisturbed at meal times for several months. If someone then enters the room while he is eating, he may start to resent anyone entering his social space for fear of losing his food and may start guarding it. When you get a puppy, you should stay in the same room as the pup while he eats. You don't have to do anything except be there and walk around so the pup becomes accustomed to a person being present while he eats without feeling threatened.

The second common reason for food guarding is when an owner is trying to show the dog who is boss and takes the dog's meal away while he is eating. Yep, that's showing him! However, dogs have limited coping strategies with situations they find unpleasant or threatening: they avoid it (not possible in this situation); resign themselves to it; or they snap. If he snaps and the person retreats (hey, it worked!) that can be the start of food guarding since the dog no longer trusts a person near him while he eats, and he has now found a way to make the person move away. Meal times are highlights of a dog's day so when he feels threatened he will react, possibly by showing aggressive behavior.

If left to his own devices, a dog will do whatever he finds emotionally or physically rewarding. If a dog finds that it's rewarding to sleep on the couch and *has been allowed to do so for some months,* he might take exception to someone suddenly denying him access to that resource and he may start to guard it. Access to the resource has probably gone un-checked or even reinforced for several months, even years. Veterinary neurologist William Klemm (1996) says, "In ways that are not yet understood, this neural origin of emotions creates internal drives or motivation that guide animals toward goal-directed behaviors." If these "goal-directed behaviors" are then reinforced, even if inadvertently by the owner, the dog will continue with the behavior. If the dog then becomes aggressive when the owner tries to remove him from the couch, the owner may misinterpret that aggression as the dog trying to raise his status rather than protecting a resource. Dogs do not understand the concept of ownership, but they are aware of resources that are rewarding. With the *right training,* an owner should be able to remove a resource the dog enjoys without the dog guarding it.

Dominance aggression

As I mentioned in Chapter 1, there is a term called 'dominance aggression,' which makes sense when considering cases where dogs act in an aggressive fashion (Bradshaw et al., 2009). This usually occurs when a dog has been physically reprimanded, bullied or has been subjected to punitive training methods and thus the dog has become fearful of his owner or some other person(s). The following are typical examples: (1) an owner comes home, having left the dog alone for some hours, and found the dog has either soiled the carpet or has chewed an item of the owner's clothing—the dog is punished by being smacked and shouted at; (2) an owner's misguided view that he has to be pack leader which entails bullying the dog; (3) first-time dog owners with little knowledge of how to raise a dog may become frustrated with a dog who is not doing what he's told, getting into mischief and end up taking their frustration out on the dog; and (4) punitive training methods like jerking the dog around at the end of a choke chain, stamping on a loose lead to get the dog to lay down, or the use of shock collars that will cause pain to the dog. All these scenarios will elicit the fear emotion—fear of their owner.

If a dog feels threatened or is being mistreated by his owner or some other person(s), he only has a limited number of strategies to cope with the situation. One option many dogs resort to is to snap or nip the owner. The owner will withdraw his hand, even momentarily, but that is sufficient to reinforce the dog's actions. The dog has now learned that using aggression works and he is likely to increase the aggression in the future. This type of aggressive problem can only be addressed by assessing the specific problem and developing a specific modification program to solve it.

The good news is that more and more dog trainers are using positive, motivational training methods such as food, toys and clicker training to reward a dog for doing something right rather than using punishment for a dog that misbehaves. These positive training methods do not rely on dominating the dog. Rather, they are more about creating a symbiotic relationship between owner and dog where the learning experience is fun for both, resulting in a better trained, happier dog. In the next chapter, I will suggest a training and management program that adheres to this philosophy.

CHAPTER 6

What's to be done?

There are now many good training books that explain how to train a dog using positive, motivational methods, and many trainers who use these methods in their classes. So, find a book and a trainer that does *not* preach pack rules and does *not* train dogs as though they are wolves. Look for books that explain how to find and raise a puppy and which have a lot of emphasis on socialization. Here are some of the key points you should consider when looking for a puppy:

- Do your homework about the breed of dog you own (or want to own).
- Select a good breeder.
- Socialize the puppy.
- Teach household etiquette and what the dog is and is not allowed to do in the home.
- Find a good training class
- Start training as early as possible using positive, motivational methods.
- Be fair to the dog.
- Be consistent in your training methods and how you treat the dog.
- Learn to read canine body language and postures.

Whilst I appreciate some people will buy a mixed-breed pup, the advice above still applies. If you know what breeds the sire and dam are then you may end up with a mix of their breed behaviors so you need to find out what those behaviors are. Some people will give a home to an adult dog from re-homing shelters in which case, some of the points will still apply.

Do your homework

It's very easy to select a breed of dog because of their looks. But looks aren't everything. Find out as much about the breed as you can before you buy one so you have a fair idea of what you're getting and be sure that he will fit in with your lifestyle. If you're not an active family, it's no good getting something like a Border Collie or a Springer Spaniel. If you don't like the idea of a lot of grooming, an Afghan Hound would not be a good choice. You wouldn't buy a car or a television just because it looks nice without knowing what you're getting, so don't buy a dog, which is a much more important purchase, without doing your homework.

Select a good breeder

Always see the dam with her litter when you visit the breeder. The reasons for seeing the dam with her litter is firstly, to make sure the person selling the puppies is actually the breeder and not an agent selling them on behalf of the breeder, and secondly, to see how the dam behaves when strangers pick up her pups. If the dam growls or acts aggressively in any way towards strangers being near or touching her pups, the pups could learn the same behavior and the last thing an owner needs is a puppy with aggressive tendencies. Ideally, you should find a breeder that raises the litter indoors rather than in an outdoors kennel. This is so the puppies can get used to the sights and sounds of household appliances, the hustle and bustle of family life and the comings and goings of different people. The sire should have been carefully selected for his temperament and other attributes the breeder is looking for, and all the hereditary health screening tests have been carried out on both sire and dam. In some cases, such as Collie Eye Anomaly or deafness, for example, the pups should be tested as well. As the socialization period starts at a very early age, make sure the breeder has started off the process. Ask the breeder what has been done in order to socialize the pup.

Socialize the puppy

The critical period of a puppy's socialization is roughly between about four and sixteen weeks. This is the period the puppy will learn that he is a dog and will shape his behavior when he's an adult. The puppy must be exposed to mild stress so he can cope with life when he's older, meet other dogs, other people and any type of stimuli likely to be faced later on. As the critical period starts so early, the breeder has an immense responsibility to start the socialization process before finding it a new home to ensure the pup has a good start in life. This is also the time to start gentle grooming and inspecting all parts of the puppy's body so he gets used to being handled. However, socialization does not stop at sixteen weeks—it's an ongoing process.

Fig. 21 and 22—Puppies should get lots of stimulation and exposure to 'scary things' during the socialization period.

Find a good training class

You should be looking for a good training class before you get your puppy home, as good ones can get booked up some weeks in advance. The ideal time to start at a puppy training class is when the pup has completed all his vaccination shots and is allowed to go out—or maybe even earlier. Some veterinarians may recommend the pup starts classes mid-way through the course of shots, but some will advise you wait until the shots are complete, so ask your vet. The class will not only help with training, but will also help with socializing your pup with people and other pups.

It's so easy and convenient to pop along to the nearest dog training class without giving too much thought about the training methods they use. If it's a training class, they must know what they are doing, right? Well, not necessarily. Many classes use training methods that are now considered outdated, like preaching the pack rules and physically jerking the dogs around at the end of a choke chain. Choke chains can cause immense physical harm to an adult dog, including; oesophageal and tracheal damage, laryngeal nerve paralysis, injury to the ocular blood vessels, and fainting to name but a few. If it can cause that much damage to an adult dog, think what might happen if used on a puppy. Never use a choke chain on a puppy. If you train the pup without a choke chain, you won't need to use one when he's grown up. So, look around for a trainer who uses positive, motivational methods of training, possibly using clicker training, but at least a class that uses food rewards or rewards of some kind that motivate your pup. Go to a class that is specifically for puppies and not one that takes dogs of all ages and mixed abilities. The class should have a limit on numbers of puppies and include a structured training program. Before you enroll, visit a class and see if the dogs and owners are happy and the instructor is using positive methods and is in control of the class. Investment in good, early training will set your dog up for life, so getting it right and going the extra mile at the beginning will pay dividends in the long term.

Some countries have an organization called the Association of Pet Dog Trainers (APDT). Their members only use kind, fair and effective methods of training, so try having a look for one in your area. For the UK, look at www.apdt.co.uk and for the US, look at www. apdt.com. If there isn't an APDT in your country, look for a class using the same techniques and ethics.

Household etiquette

Having brought a well-adjusted puppy home, start teaching him what he is allowed to do in the home and what he is not allowed to do. If you don't want a dog lying on your sofa, don't encourage him to do so when he's a puppy. Alternatively, teach a command to invite your puppy onto the sofa, so you can have the company of your dog when you like, but also teach an 'off' command so he will get off when you ask. If you don't want a dog jumping up on people, don't encourage it as a puppy.

Fig. 23—Oh well. You can't win them all!

Be with the puppy when he is eating. Even stroke the puppy while he is eating so he becomes comfortable with people being around at meal times. Add a little more food to the bowl while he is eating so he learns that a hand by his food bowl is a good thing.

Play games of tug and even let the puppy win sometimes. The puppy should perceive games of tug as just that, a game, but don't forget to teach a 'drop' behavior.

Start training early

It is essential to have a well-mannered dog and very basic training can start when the pup is just a few weeks old. A pup's brain grows to 50% of its full size by the time it's just two months old, then to 80% at four months and fully grown by twelve months (Coppinger and Coppinger 2001), so training can and should start very early in the pup's life. Socialization and basic obedience go hand-in-hand. Start at home before the pup has completed his vaccinations and then take your puppy to a training class specifically for puppies so he can learn to interact with puppies of a similar age. Use reward-based, motivational methods of training so the pup finds learning rewarding. The more rewarding, the more he's likely to repeat the behavior. Include in the training the commands 'drop' and 'leave it' which should be taught to a puppy. Unfortunately, these commands are usually taught to an adolescent dog when the owner realizes how important those commands are when the dog is running around the garden with a pair of the owner's underpants! If you sometimes want your dog to be on the sofa with you, teach commands to invite him onto the sofa, and to ask him to get off when you say.

Be fair to the dog

Do not expect the puppy to instantly be the perfect household pet. Teaching house etiquette and basic obedience will take time, so be patient. Dogs can only think and act in canine ways so it will take time for him to learn what we expect of him using our ways of communication. Dogs are happy to live in a social structure. If a structure doesn't exist, the dog will just do whatever he finds rewarding. Do not physically punish your puppy for any wrongdoing. Chances are it was your fault anyway!

Be consistent

Lack of consistency with commands and how the dog is treated is a very common problem, and many owners are not even aware that they are being inconsistent in terms of what they want from their dog. It is very easy to confuse a dog if he is given conflicting commands that are supposed to mean the same thing or he is allowed to do one thing with one member of the family but not another. This may result in the dog developing unwanted behaviors due to stress and conflicting emotions. So be consistent in your commands and how you treat the dog.

Learn to read canine body language

We expect dogs to understand our verbal commands and very often visual cues. It's only right, therefore, that we learn to understand what a dog is trying to tell us by learning to 'read' his body posture and facial expression. We will then be in a better position to understand what our dog is trying to tell us—whether he's happy, sad, in pain, needs to be let out to eliminate, at rest or perhaps there may be the onset of a slight problem which could escalate into a major problem if the signs are ignored. A typical example of this is when a dog is left on his own when he's having his meals as mentioned above. If this continues for several months, he may become anxious when someone does come into the room when he's eating. A dog who is content to have people in the same room will have his tail in its normal, relaxed position. If there is a slight sway of the tail or it is tucked slightly between his legs, then the pup is showing signs of anxiety and this is the time to take remedial action.

Fig. 24—Relaxed tail position when eating—no problem!

Some dogs will come from a rescue organization or shelter. Although the principles of what I've said will apply to older dogs, re-homed dogs may already have behavioral problems that will need treatment. They may lack training or they may need special treatment if it's a case of neglect, so older, re-homed dogs must be trained very carefully.

CHAPTER 7

Conclusions

The idea of a domestic dog exhibiting 'dominance' over his owner originated from observing the behavior of captive wolf packs. However, research by Mech and others shows that a free pack of wolves does not have a dictatorial alpha wolf ruling over the pack. These findings show dominance challenges between wolves are rare, if they exist at all; instead they act as a family, sharing labor. *If* we were to apply the wolf's behavior to our dogs, then it's the behavior from the free packs that we should be looking at and not the observed behavior of a captive pack that has a more competitive lifestyle.

While dogs evolved from wolves, clearly they are now significantly different animals both physically and behaviorally. Yet, the 'pack rules' we are told to use on our dogs are supposed to reflect wolf behavior, which in large part, they don't. Something doesn't quite fall into place here. To add to the confusion, pack rules are based on canine-to-canine communication, but we can only communicate using human methods—we cannot mimic canine behavior any more than a dog can mimic ours. Dogs are conspecific so they won't perceive themselves as part of a human pack, therefore if we try to enforce pack rules, the dog won't understand what we are trying to do and we could end up with one very confused, depressed dog.

As we provide all the elements of survival and comfort for our dog, why would he need to raise his status?

Accepting that the dog has evolved from the wolf and is now either a different species or a sub-species (depending on which canine authority you read), we can start to treat him as a dog and not as a wolf in dog's clothing. The dog is not going to be dominant and raise his status in the human pack because he's not part of the human pack. Part of our social unit? Absolutely—but not our pack. Equally, we don't have to be dominant over the dog by using so-called pack rules as they are totally inappropriate and the dog won't understand what we are trying to tell him.

There are two aspects of dominance that do apply to pet dogs. The first is when a dog, either through unfettered access or inadvertent training, learns how to maintain access or control over a particular resource, be it a food bowl, toy or couch. In that case, the dog may display what can be termed resource guarding to maintain control over a resource from other dogs or people. Fortunately, that aspect of dominance can be avoided or solved though proper training and management techniques that do not involve the use of aversive methods.

The second aspect of dominance that is relevant is when a 'dominant' owner uses aversive or harsh methods to teach or train a dog in order to get their own way or uses physical abuse for whatever reason. This results in a suppressed dog not being able to truly express his feelings. This can lead to learned aggression as the dog attempts to defend himself. Rather than a relationship built on firm bonds and mutual understanding, the relationship turns destructive.

We have chosen to domesticate the dog and we therefore have a responsibility to understand the dog as best we can. We are both social species which is why we can co-exist together. Dogs have to exist on our terms, but we need to be fair, consistent and understanding to their needs.

We should have a symbiotic relationship with our dogs—that is, we live together to our mutual benefit. We provide our dog with everything he could possibly need. We in return enjoy the experience of owning a dog because we like dogs—we enjoy their company, we enjoy participating in dog sports and going on long walks with them. If we can accept that our domestic dog does not perceive itself as part of our human pack, we can then start to treat a dog for what he actually is—a dog.

Personally, I don't like putting labels on what we think we are in our relationship with our dogs, but some people do. Clearly thinking of ourselves as the 'alpha' is outdated and incorrect. 'Pack leader' is also a misnomer as domestic dogs cannot form packs in the true sense of the word, so there isn't a pack to lead. And even if dogs could form packs, why would they form one with a different species, i.e., us? And why use the word 'dominant' just because we are talking about our dogs? Parents are dominant over their children, but they don't enforce silly rules. Sensible rules, yes, to help and guide them, so why can't we do the same with our dogs? After all, they are part of our social unit. If I were to use a label, it would be 'responsible dog owner'—the emphasis on 'responsible' and all that goes with it. We don't have to be alpha, dominant or pack leader. All we need to be is an owner responsible for guiding his dog and influencing his behavior through socializing and training him to live in harmony within our society. We also owe it to our dog to learn about dog behavior so we can understand better our canine companion. If we follow *those* rules, we should have no fear of dogs taking over our family.

BIBLIOGRAPHY

Abrantes R. (1997) *The Evolution of Canine Social Behavior*, Wakan Tanka Publishers p. 69, 70.

Boitani L, Ciucci P, Ortolani A, (2008) *The Behavioural Biology of Dogs* (ed P. Jensen) CAB International, p.153, 154.

Bradshaw J W S, Blackwell, E. J, Casey R A, (2009) "Dominance in domestic dogs–useful construct or bad habit?," *Journal of Veterinary Behaviour*, 4, 3, 140, 143.

Clutton-Brock J. (1999) *The Domestic Dog* (ed. Serpell J.) Cambridge University Press p. 9, 15.

Coppinger R. & Coppinger L. (2007) *Think Ethology Unit 7*, Centre of Applied Pet Ethology (www.coape.org).

Coppinger R. & Coppinger L. (2001) *Dogs. A Startling New Understanding of Canine Origin, Behavior & Evolution*, Scribner, pp 51–61, 67, 81, 112, 206, 209, 210.

Donaldson J. (1996) *The Culture Clash*, James & Kenneth Publishers, p. 19.

Dunbar I. (1979) *Dog Behavior*, TFH Publications, p. 84.

Dunbar I. (2006) *San Francisco Chronicle*, October 15, 2006.

Feddersen-Petersen D. U. (2008) *The Behavioural Biology of Dogs*, (ed P. Jensen) CAB International, p.111-112.

Fisher J. (1997) *Diary of a Dotty Dog Doctor*, (ed. Whitehead S.) Alpha Publishing, p. 106.

Fisher J. and Whitehead S. (2001) *Advanced Think Dog,* course units.

Goodwin D, Bradshaw J W S, Wicken S M, (1997) "Paedomorphosis affects agonistic visual signals of domestic dogs," *Animal Behavior,* 53, 297–304.

Harrington F. and Asa C. (2003) *Wolves–Behavior, Ecology, and Conservation* (ed. Mech D. & Boitani L.), University of Chicago Press p.93, 94.

Klemm W. (1996) *Understanding Neuroscience,* Mosby-Year Book Inc. p. 198.

Kreeger T. (2003) *Wolves–Behavior, Ecology, and Conservation* (ed. Mech D. & Boitani L.), University of Chicago Press, p. 192, 193.

Lindsay S. (2000) *Handbook of Applied Behavior and Training Vol. 1,* Blackwell Publishing, p. 12, 104, 381.

McGreevy P., McLean N. (2009) "Punishment in horse-training and the concept of ethical equitation," *Journal of Veterinary Behavior,* 4 5 195.

Mech. L. D. (1999) "Alpha, status, dominance, and division of labor in wolf packs." *Canadian Journal of Zoology,* 77 (8) (1999), p. 1198, 1200.

Mech. L. D. (2000) "Leadership in Wolf, Canis lupus, Packs. Canadian Field-Naturalist" 114(2):259-263. Jamestown, ND: Northern Prairie Wildlife Research Center Online. See: www.npwrc.usgs.gov/resource/mammals/leader/index.htm (Last viewed 2 July 2010).

Mech L.D. (2003) *The Wolf. The Ecology and Behavior of an Endangered Species,* University of Minnesota Press, p. 4, 7, 49, 75, 76, 121, 129, 155, 185.

Mech L. D. (2008), "Whatever Happened to the Term Alpha?" *International Wolf Magazine,* Winter, 2008 P 4-8 http://www.wolf.org/wolves/news/pdf/winter2008.pdf (Last viewed 2 July 2010).

Overall K. (1997) *Clinical Behavioral Medicine for Small Animals,* Mosby-Year Book Inc., p. 115, 512.

Overall K. (2003) US APDT Conference, Orlando.

Packard M. (2003) *Wolves–Behavior, Ecology, and Conservation* (ed. Mech D. & Boitani L.) University of Chicago Press, p.40, 42.

Penman S. (1994) *Veterinary Notes for Dog Owners* (ed. Turner T.), Popular Dogs Publishing, p. 489.

Pryor K. (2002) *Don't Shoot the Dog,* Ringpress Books, p. 108.

Rooney N.J., Bradshaw J.W.S., Robinson I. H. (2000) "A comparison of dog–dog and dog–human play behavior," *Applied Animal Behavior Science* 66 (2000) 235–248.

Udell M.A. R., Wynne C. D. L. (2008) "A review of domestic dogs (Canis Familiaris) human-like behaviours or why a behavioural analysts should stop worrying and love their dogs," *Journal of the Experimental Analysis of Behavior,* 2008, 89, 253.

Semyonova A. (2003) *The social organization of the domestic dog; a longitudinal study of domestic canine behavior and the ontogeny of domestic canine social systems.* The Carriage House Foundation, The Hague, Netherlands. www.nonlineardogs.com (2006 version), p.2, 33 (Last viewed 2 July 2010).

Schmidt P.A. and Mech L.D. (1997) "Wolf Pack Size and Food Acquisition," *American Naturalist* 150 (4):513-517. Jamestown ND: Northern Prairie Wildlife Research Center Online. http://www.npwrc.usgs.gov/resource/mammals/wpsize/index.htm (Last viewed 2 July 2010).

Zimen E. (1983) *The Wolf: His Place in the Natural World,* Souvenir Press, p. 26, 27, 30, 173

About the Author

Barry Eaton has a Diploma in Companion Animal Behaviour and Training (COAPE NOCN) and an Advanced Cert. in 'Introduction to Ethology.' He is an Affiliate of COAPE and a Member of the CAPBT (www.capbt.org). He is an experienced dog trainer and keeps abreast of scientific research into canine behaviour. He specializes in training dogs that are deaf. He has written a book, *Hear, Hear* (www.deaf-dogs-help.co.uk) which provides help on how to train dogs that are born deaf. Barry has acted as consultant for Usborne Publishing on three of their books about dogs and contributed to the UK Association of Pet Dog Trainers, 'Teach Yourself Dog Training.'

Barry's experience includes appearing on TV and radio, lecturing widely in the UK and Europe, and as a former Chair of the Wessex Sheepdog Society, enjoyed successfully participating in sheep dog trials.

INDEX

From Dogwise Publishing
www.dogwise.com · 1-800-776-2665

BEHAVIOR & TRAINING

ABC's of Behavior Shaping. Proactive Behavior Mgmt, DVD set. Ted Turner

Aggression In Dogs. Practical Mgmt, Prevention, & Behaviour Modification. Brenda Aloff

Am I Safe? DVD. Sarah Kalnajs

Barking. The Sound of a Language. Turid Rugaas

Behavior Problems in Dogs, 3rd ed. William Campbell

Brenda Aloff's Fundamentals: Foundation Training for Every Dog, DVD. Brenda Aloff

Bringing Light to Shadow. A Dog Trainer's Diary. Pam Dennison

Canine Body Language. A Photographic Guide to the Native Language of Dogs. Brenda Aloff

Changing People Changing Dogs. Positive Solutions for Difficult Dogs. Rev. Dee Ganley

Chill Out Fido! How to Calm Your Dog. Nan Arthur

Clicked Retriever. Lana Mitchell

Do Over Dogs. Give Your Dog a Second Chance for a First Class Life. Pat Miller

Dog Behavior Problems. The Counselor's Handbook. William Campbell

Dog Friendly Gardens, Garden Friendly Dogs. Cheryl Smith

Dog Language, An Encyclopedia of Canine Behavior. Roger Abrantes

Dogs are from Neptune. Jean Donaldson

Evolution of Canine Social Behavior, 2nd ed. Roger Abrantes

From Hoofbeats to Dogsteps. A Life of Listening to and Learning from Animals. Rachel Page Elliott

Get Connected With Your Dog, book with DVD. Brenda Aloff

Give Them a Scalpel and They Will Dissect a Kiss, DVD. Ian Dunbar

Guide to Professional Dog Walking And Home Boarding. Dianne Eibner

Language of Dogs, DVD. Sarah Kalnajs

Mastering Variable Surface Tracking, Component Tracking (2 bk set). Ed Presnall

My Dog Pulls. What Do I Do? Turid Rugaas

New Knowledge of Dog Behavior (reprint). Clarence Pfaffenberger

Oh Behave! Dogs from Pavlov to Premack to Pinker. Jean Donaldson

On Talking Terms with Dogs. Calming Signals, 2nd edition. Turid Rugaas

On Talking Terms with Dogs. What Your Dog Tells You, DVD. Turid Rugaas

Play With Your Dog. Pat Miller

Positive Perspectives. Love Your Dog, Train Your Dog. Pat Miller

Positive Perspectives 2. Know Your Dog, Train Your Dog. Pat Miller

Predation and Family Dogs, DVD. Jean Donaldson

Quick Clicks, 2nd Edition. Mandy Book and Cheryl Smith

Really Reliable Recall. Train Your Dog to Come When Called, DVD. Leslie Nelson

Right on Target. Taking Dog Training to a New Level. Mandy Book & Cheryl Smith

Stress in Dogs. Martina Scholz & Clarissa von Reinhardt

Tales of Two Species. Essays on Loving and Living With Dogs. Patricia McConnell

The Dog Trainer's Resource. The APDT Chronicle of the Dog Collection. Mychelle Blake (*ed*)

The Dog Trainer's Resource 2. The APDT Chronicle of the Dog Collection. Mychelle Blake (*ed*)

The Thinking Dog. Crossover to Clicker Training. Gail Fisher

Therapy Dogs. Training Your Dog To Reach Others. Kathy Diamond Davis

The Toolbox for Building a Great Family Dog. Terry Ryan

Training Dogs. A Manual (reprint). Konrad Most

Training the Disaster Search Dog. Shirley Hammond

Try Tracking. The Puppy Tracking Primer. Carolyn Krause

Visiting the Dog Park, Having Fun, and Staying Safe. Cheryl S. Smith

When Pigs Fly. Train Your Impossible Dog. Jane Killion

Winning Team. A Guidebook for Junior Showmanship. Gail Haynes

Working Dogs (reprint). Elliot Humphrey & Lucien Warner

HEALTH & ANATOMY, SHOWING
Advanced Canine Reproduction and Whelping. Sylvia Smart

An Eye for a Dog. Illustrated Guide to Judging Purebred Dogs. Robert Cole

Annie On Dogs! Ann Rogers Clark

Another Piece of the Puzzle. Pat Hastings

Canine Cineradiography DVD. Rachel Page Elliott

Canine Massage. A Complete Reference Manual. Jean-Pierre Hourdebaigt

Canine Terminology (reprint). Harold Spira

Breeders Professional Secrets. Ethical Breeding Practices. Sylvia Smart

Dog In Action (reprint). Macdowell Lyon

Dog Show Judging. The Good, the Bad, and the Ugly. Chris Walkowicz

Dogsteps DVD. Rachel Page Elliott

The Healthy Way to Stretch Your Dog. A Physical Therapy Approach. Sasha Foster and Ashley Foster

The History and Management of the Mastiff. Elizabeth Baxter & Pat Hoffman

Performance Dog Nutrition. Optimize Performance With Nutrition. Jocelynn Jacobs

Positive Training for Show Dogs. Building a Relationship for Success Vicki Ronchette

Puppy Intensive Care. A Breeder's Guide To Care Of Newborn Puppies. Myra Savant Harris

Raw Dog Food. Make It Easy for You and Your Dog. Carina MacDonald

Raw Meaty Bones. Tom Lonsdale

Shock to the System. The Facts About Animal Vaccination... Catherine O'Driscoll

Tricks of the Trade. From Best of Intentions to Best in Show, Rev. Ed. Pat Hastings

Work Wonders. Feed Your Dog Raw Meaty Bones. Tom Lonsdale

Whelping Healthy Puppies, DVD. Sylvia Smart

Phone in your Order! 1.800.776.2665 8am-4pm PST / 11am-7pm EST

Sign in | View Cart

Be the First to Hear the News!
Have New Product and Promotion
Announcements Emailed to You.
Click Here To Sign Up!

Free Shipping for Orders over $75 - click here for more information!

Win a $25 Dogwise credit - click here to find out how!

Featured New Titles

STRESS IN DOGS - LEARN HOW DOGS SHOW STRESS AND WHAT YOU CAN DO TO HELP, by Martina Scholz & Clarissa von Reinhardt
Item: DTB909
Is stress causing your dog's behavior problems? Research shows that as with humans, many behavioral problems in dogs are stress-related. Learn how to recognize when your dog is stressed, what factors cause stress in dogs, and strategies you can utilize in training and in your daily life with your dog to reduce stress.
Price: $14.95 more information...

SUCCESS IS IN THE PROOFING - A GUIDE FOR CREATIVE AND EFFECTIVE TRAINING, by Debby Quigley & Judy Ramsey
Item: DTO230
The success is indeed in the proofing! Proofing is an essential part of training, but one that is often overlooked or not worked on enough. We all know the story of the dog who can perform a variety of behaviors perfectly in the backyard but falls apart in the obedience ring. This book is full of great ideas and strategies to help your dog do his best no matter what the distractions or conditions may be. Whether competing in Rally or Obedience, trainers everywhere will find this very portable and user friendly book an indispensable addition to their tool box.
Price: $19.95 more information...

REALLY RELIABLE RECALL DVD, by Leslie Nelson
Item: DTB810P
From well-known trainer Leslie Nelson! Easy to follow steps to train your dog to come when it really counts. In an emergency. Extra chapters for difficult to train breeds and training class instructors.
Price: $29.95 more information...

THE DOG TRAINERS RESOURCE - APDT CHRONICLE OF THE DOG COLLECTION, by Mychelle Blake, Editor
Item: DTB880
The modern professional dog trainer needs to develop expertise in a wide variety of fields: learning theory, training techniques, classroom strategies, marketing, community relations, and business development and management. This collection of articles from APDT's Chronicle of the Dog will prove a valuable resource for trainers and would-be trainers.
Price: $24.95 more information...

SHAPING SUCCESS - THE EDUCATION OF AN UNLIKELY CHAMPION, by Susan Garrett
Item: DTA260
Written by one of the world's best dog trainers, Shaping Success gives an excellent explanation of the theory behind animal learning as Susan Garrett trains a high-energy Border Collie puppy to be an agility champion. Buzzy's story both entertains and demonstrates how to apply some of the most up-to-date dog training methods in the real world. Clicker training!
Price: $24.95 more information...

FOR THE LOVE OF A DOG - UNDERSTANDING EMOTION IN YOU AND YOUR BEST FRIEND, by Patricia McConnell
Item: DTB890
Sure to be another bestseller, Trish McConnell's latest book takes a look at canine emotions and body language. Like all her books, this one is written in a way that the average dog owner can follow but brings the latest scientific information that trainers and dog enthusiasts can use.
Price: $24.95 more information...

HELP FOR YOUR FEARFUL DOG: A STEP-BY-STEP GUIDE TO HELPING YOUR DOG CONQUER HIS FEARS, by Nicole Wilde
Item: DTB878
From popular author and trainer Nicole Wilde! A comprehensive guide to the treatment of canine anxiety, fears, and phobias. Chock full of photographs and illustrations and written in a down-to-earth, humorous style.
Price: $24.95 more information...

FAMILY FRIENDLY DOG TRAINING - A SIX WEEK PROGRAM FOR YOU AND YOUR DOG, by Patricia McConnell & Aimee Moore
Item: DTB917
A six-week program to get people and dogs off on the right paw! Includes trouble-shooting tips for what to do when your dog doesn't respond as expected. This is a book that many trainers will want their students to read.
Price: $11.95 more information...

THE LANGUAGE OF DOGS - UNDERSTANDING CANINE BODY LANGUAGE AND OTHER COMMUNICATION SIGNALS DVD SET, by Sarah Kalnajs
Item: DTB875P
Features a presentation and extensive footage of a variety of breeds showing hundreds of examples of canine behavior and body language. Perfect for dog owners or anyone who handles dogs or encounters them regularly while on the job.
Price: $39.95 more information...

THE FAMILY IN DOG BEHAVIOR CONSULTING, by Lynn Hoover
Item: DTB887
Sometimes, no matter how good a trainer or behavior consultant you are, there are issues going on within a human family that you need to be aware of to solve behavior or training problems with dogs. For animal behavior consultants, this text opens up new vistas of challenge and opportunity, dealing with the intense and sometimes complicated nature of relationships between families and dogs.
Price: $24.95 more information...